# A
# Relaxed
# Life


## By

## Isaac Lasley

# A Relaxed Life

## Isaac Lasley

## Lulu.com Edition

ISBN: 978-1-387-91852-2

This book is written to help people who read it to live better, more fulfilling lives. They will hopefully be free from pains and ailments that plague the lives of many people of the world. They can take comfort in the facts that even the worst conditions, like cancer; have many natural treatments that can possibly be more effective than medical treatments. Natural remedies existed before modern medicine, and many of the natural remedies work better.

# Contents

# Introduction

Herbal supplements and complementary treatment may be used by anybody of any health to improve their personal disposition and to physically feel better on a day to day basis. There are supplements that will naturally help with unhealthy bowels, lowering cholesterol, muscle pain, joint pain, swelling/inflammation, diarrhea, a laxative, circulatory problems, eye irritation, skin conditions, ulcers, frequent urination, appetite stimulus, enlarged prostate, premenstrual syndrome, rheumatoid arthritis, stomach ache, tooth ache, respiratory problems, pressure behind eyes, heart disease, nausea, oral pain, back pain, improve heart rate, high blood pressure, mouth sores from cancer treatment, menstrual cramps, menopausal symptoms, night sweats, hot flashes, heal external injuries, to help prevent cancer, help fight cancer, help fight and prevent viruses, improve energy, help chronic conditions, sooth skin irritation, help with weight loss, help improve body's homeostasis, regulate menstruation, help prevent disease, improve immune system, reduce fatigue, reduce effects of cancer and treatment, reduce chemo side effects, prevent bone loss or fragile bones, improve mood, improve sleep quality, improve immune system, improve circulation, improve thought processes, improved mental alertness, improve liver functions, and improve digestion.

There is complementary therapy for back pain, fatigue, swelling/inflammation, migraine headaches, chronic pain, allergies, nausea, physical stress relief, high blood pressure, tension, respiratory problems, insomnia, fungal growth, bacteria problems, to improve bodily functions/processes, control weight, improve immune system function, slow circulation, improve circulation, improve stamina, reduce swelling, alleviate aches and pains, to help build muscle, physically feel better, improve body's energy flow, gain harmony with world and universe around you, to improve social life, psychologically feel better, beat depression, improve mood, relieve anxiety, increase will to live, better deal with emotional stress, relive mental tension, improve general temperament, obtain peace of mind, improve quality of life, gain greater self expression, increase self awareness, gain better personal disposition, improve personal life outlook, come to terms with emotional conflict, improve body's personal balance; as well as helping to improve mind, body, and spirit.

# Herbal Supplements

You may want to consult your doctor before beginning to use supplements to help avoid any possible negative interactions with drugs or treatments you are already using. Specific herbal supplements may include: acidophilus, aloe, antioxidants, arnica, astralagus, bilberry, black cohosh, bromelain, calcium, canarius, capsicum, carnitine, castor oil, chasteberry, Chinese thunder god vine, chlorella, clove supplements, comfrey, copper, ellagic acid, feverfew, flaxseed, folate, garlic, ginger, gingko, ginseng, gotu kola, grapes/seed extract, green tea, hawthorn, horse chestnut, Indian snakeroot, kava, lavender, licorice root extract, lycopene, maitake mushroom, marijuana, melatonin, milk thistle, modified citrus pectin, noni plant, omega-3 fatty acids, peppermint oil, pokeweed, psyllium, rabdosia rubescens, sea vegetables, shark liver oil, shiitake mushroom, soy, St. John's wort, tea tree oil, turmeric, valerian root, vitamin C, and/or vitamin D.

*Acidophilus* is a healthful bacterium that is commonly found in the digestive tracts of mammals and is also found in many dairy products such as yogurt. This supplement is promoted to help to keep bowels healthy, lower cholesterol, and possibly may help to prevent cancer [1].

*Aloe* is a gel from the aloe vera plant that is commonly used to sooth skin conditions, while the juice/latex from the plant is utilized for constipation [2, 3].

*Antioxidants* may be noted as vitamin A, vitamin C, vitamin E, beta-carotene, lutein, and lycopene. While in the past these nutrients were merely looked upon as vitamins that help to prevent cancer occurrence, science is less sure of that statement now due to conflicting research. These nutrients do help the body to maintain homeostasis and prevent common disease by destroying molecules in the body known as "free radicals", which are activated oxygen molecules that have the potential to cause disease [4].

*Arnica* is an herb that is commonly applied to the skin in order to treat inflammation, wounds, and infections, as it is promoted for skin irritation, muscular/joint pain, and to heal exterior tissue injuries [5].

*Astralagus* is an herbal root that has been used by Chinese herbalists for thousands of years to help enhance the functioning of the immune system and help the body to increase its energy in an effort to prevent disease. This herb may be helpful to reduce the effects of cancer and the side effects of cancer treatment, as well as fatigue [6, 7].

*Bilberry* is an herb that is commonly used to treat symptoms such as diarrhea, circulatory problems, and menstrual cramps [8].

*Black cohosh* roots are commonly used to help women to find relief from menopausal symptoms such as night sweats and hot flashes, as well as some others [9, 10, 11].

*Bromelain* is an enzyme that is commonly found in pineapples which helps to reduce inflammation as well as swelling in soft tissue, and may help to ease the side effects of chemotherapy [12].

*Calcium* supplements may be beneficial to cancer victims in order to help prevent loss of bone mass, fragile bones, osteoporosis, and possibly some cancer types [13, 14].

*Canarius* ("Yerbe Mate") is an herb in the basil family that helps to naturally cleanse the body of impurities while on some level also promotes a meditative and reflective state of mind. It does this by increasing the oxygen flow to all systems of the body by improving the way red blood cells absorb oxygen from the lungs and distribute it to all systems of the human body. The white blood cell count in the body is also likely to increase and help to more effectively remove foreign particles within the body, as well as to help the body more easily repair damaged tissues. The most beneficial way to utilize canarius is by adding a tea spoon about two thirds of the way full in a common sized tea cup, then filling the cup with boiling water. To be effective it should be left to steep with a saucer or something to cover the cup and lock in the steam for about ten to fifteen minutes at least. You can also add a tablespoon , or two, to your iced tea pitcher and let it sit overnight in the fridge for positive effects. A little goes a long way with this herb and it becomes more potent as it ferments.

*Capsicum* supplements are most commonly used to treat post-surgical pain; and to provide temporary relief from mouth sores resulting from radiation therapy or chemotherapy [15].

*Carnitine* is found in nearly all the cells in the body and deficiency of this amino acid can result in fatigue. Carnitine supplements may help victims of cancer to reduce their fatigue, as well as improve their mood and likely help to better their quality of sleep [16].

*Castor oil* supplements are extracted from ricinus communis seeds, an herb from India and Africa, as it has been used since the times of ancient Egypt internally as a laxative, and externally to treat eye irritation, skin conditions, or to deliver chemotherapy drugs to malignant tumors [17].

*Chasteberry* is supplement that has been used for thousands of years to treat menopausal symptoms and premenstrual syndrome [18].

*Chinese thunder god vine* root has been used for hundreds of years to suppress an overactive immune system, as well as, treat inflammation, excessive menstruation, and rheumatoid arthritis, among other diseases. Thunder god vine may also have anti-cancer properties and effects as a result of its use [19].

*Chlorella* supplements are single celled freshwater alga from Japan that contains antioxidants which may help to inhibit the growth of cancer cells, although the anti-cancer properties have not been proven beyond doubt at this point [20].

*Clove supplements*, used in China for hundreds of years, can be noted to relieve oral pain, and may be an antioxidant as it has been shown to block free radicals within the body. With that in mind this supplement may have anti-cancer properties that are yet unnoted [21].

*Comfrey* roots and leaves have been used for many years to make a cream that helps to treat physical injuries, swelling, arthritis, muscular aches, and back pain as it is applied externally [22].

*Copper* is found naturally in the body as it helps to regulate heart rate and blood pressure. Copper may help to regulate antioxidants which help the immune system showing anti-cancer properties, but high levels of copper in the body may contribute to cancer cell formation as copper is needed to form new blood vessels [23].

*Ellagic acid* supplements have only been tested in laboratory and animal tests, but have shown very promising results as it has been shown to cause the death of cancer cells, but this has not been verified in human subjects [24].

*Feverfew* has been used for hundreds of years to treat aches and pains like headaches, toothaches, and stomach aches [25].

*Flaxseed* and its oil are commonly used to help with troubles such as constipation, respiratory problems, menstrual problems, high cholesterol, and arthritis, as flaxseed/oil may have anti-cancer properties [26].

*Folate (folic acid)* is naturally occurring water soluble B6 that is involved in our DNA (genetic map) function, repair, and synthesis. A deficiency in folate may contribute to an increased risk of developing cancer, where as regular intake of folic acid from supplements or food may result in a lower risk of developing cancer [27, 28].

*Garlic* is an herbal supplement that is commonly used for high blood pressure, high cholesterol, and heart disease. Research does show that people who tend to regularly eat more garlic in their balanced diet tend to have a lower risk for developing certain cancers. Garlic may also reduce tumor growth, as some laboratory studies have shown that it may help to naturally kill cancer cells [29, 30, 31].

*Ginger* is commonly used to reduce the sickness, or cancer, symptom/treatment side effect of nausea and as a result lessen vomiting as well [32, 33, 34]. Maybe, enjoy sipping a ginger ale when you're a bit sick.

*Gingko* has been promoted as a memory enhancer for years, but it merely improves blood circulation and as a result boosts the oxygen level in the brain as well as throughout the body, resulting in enhanced functioning. Of course, enhanced blood flow essentially means that gingko is basically a blood thinner [35, 36].

*Ginseng* has been used as a part of Chinese medicine for thousands of years as it contains chemicals known as ginsenocides, which are thought to be responsible for ginseng's medicinal properties as this herb has been shown to enhance energy and may improve mental cognition. Ginseng may also have anti-cancer properties since some victims of cancer have shown a slower growth rate and longer life as a result in correlation with taking ginseng supplements regularly before and/or after cancer is diagnosed [37, 38].

*Gotu kola* has been used for thousands of years by many cultures around the world to help improve circulation and reduce inflammation, in addition to many other ailments. It has also been used to improve memory, promote relaxation, and aid in meditation [39].

*Grapes/seed extract* contains powerful antioxidants which are beneficial to the immune system, as grapes have been positively associated for centuries. While grapes and grape seed extract are utilized to help patients to deal with high blood pressure, circulatory problems, and high cholesterol for many years, this supplement may also have anti-cancer properties as a result of the powerful antioxidants, although science is unsure of that [40, 41].

*Green tea* is created from the dried and steamed leaves of the camellia sinesis shrub that is native to Asia, as this herb contains polyphenol chemicals which have antioxidant properties. Green tea has been used for millennia to treat digestive and cardiovascular problems, as well as to prevent infections, lower cholesterol, help with weight loss, and to improve mental alertness. This supplement is now also used to help prevent the development of cancer, as well as to help treat many cancers, even though the scientific studies that have been performed have shown mixed results [42, 43, 44].

*Hawthorn* is commonly used to help individuals to improve the functioning of their heart and aid to help prevent heart failure [45].

*Horse chestnut* is commonly used to help treat chronic venous insufficiency, which can be noted by difficulty with veins efficiently returning blood to the heart from extremities, especially the legs [46].

*Indian snakeroot* has been used for thousands of years in Asia, as it is commonly utilized as a sedative for relaxation or meditation and may also be used to treat high blood pressure. While science is unsure, this Indian snakeroot contains chemicals that may have anti-cancer properties [47].

*Kava* is commonly used to promote relaxation, reduce anxiety, and is often utilized as a sleep aid. As a topical agent applied directly to the skin, kava can be noted as a local anesthetic [48, 49].

*Lavender* is commonly used to help treat hair loss, insomnia, nausea, headache, anxiety, and depression as it may be used during aromatherapy or taken orally [50].

*Licorice root extract* has been utilized by the Chinese culture for millennia, as this herbal extract, among other things, helps to promote the healing of peptic ulcers and respiratory viral infections. Science has also noted that licorice may have anti-cancer properties as it has been found to help prevent DNA mutation, inhibit the formation of tumors, and kill cancer cells [51, 52].

*Lycopene* is an antioxidant, most commonly found in tomatoes as well as many other fruits and vegetables which gives them their color, has been noted as a great supplement, since individuals with regular tomato/lycopene intake tend to show a lower risk for many cancers. As an antioxidant lycopene may also have anti-cancer properties since it may help to suppress the spread and slow the growth of tumors [53].

*Maitake mushroom*s, common to east Asia, contains the polysaccharide, beta glucan (beta glycan), which has been shown to help stimulate the immune system, as well as activate proteins and cells such as T-cells and macrophages, among others that attack cancer cells which may result in slowed tumor growth or possibly remission [54].

*Marijuana*, scientifically known as cannabis sativa, grows throughout the world, but is more common to warmer tropical climates and contains the drug tetrahydracannibol (THC). The buds and leaves of this plant have been used for thousands of years by cultures throughout the world as an effective herbal remedy for nausea, vomiting, pain, pressure behind the eyes, anxiety, insomnia, and as an appetite stimuli [55, 56].

*Melatonin* is commonly used as a sleep aid, but may also have properties associated with cancer as individuals who tend to show low levels of melatonin in their system may have a higher risk of developing certain types of cancer. Melatonin may have anti-cancer properties as well, since some cancer growth has been shown to be

16

slowed, as various other cancers went into partial or total remission as a result of its use [57, 58].

*Milk thistle* has been used for thousands of years since it contains the antioxidant silymarin, as it is most commonly used to help improve the functioning of the liver, but may also be beneficial to help lower cholesterol and slow the growth of some cancers [59, 60, 61].

*Modified citrus pectin* is a carbohydrate supplement that may have anti-cancer properties as it may help to prevent metastasis of some cancers [62].

*Noni plant* supplements may be used to help skin conditions, joint pain, and chronic conditions. This supplement may also have anti-cancer properties as some studies have shown it to help prevent tumor growth, metastasis, and it may also help to prevent normal cells from developing into cancer cells [63, 64].

*Omega-3 fatty acids*, include alpha-linolenic acid that comes from foods such as olives/oil, some beans, canola, flaxseed/linseed, and english walnuts; as well as eicosapentaenoic acid and docosahexaenoic acid which both are present in fish and fish oil supplements. Omega-3 fatty acids are most commonly used to help to reduce high blood pressure, prevent heart disease, as well as reduce the risk of other cardiovascular disorders, but may also help to slow the growth of certain types of cancer. It is unclear whether this supplement may help to prevent cancer, and should not be taken with such expectation [65, 66, 67, 68].

*Peppermint oil* is commonly used to reduce nausea as well as other minor digestive system problems [69, 70]. *Peppermint tea* can also help minor respiratory ailments or sickness.

*Pokeweed* is still having research conducted on it, but it has shown promising results in animal studies, as it contains a protein noted as pokeweed antiviral protein (PAP). PAP may have anti-cancer properties as it definitely has anti-viral properties since it has been shown to effectively take action against viruses such as herpes and HIV. This Native American remedy appears to help increase the overall effectiveness of the immune system [71].

*Psyllium* is common to Asia and Mediterranean Europe and it has commonly been used for generations to effectively treat constipation, in addition to reducing cholesterol when taken with abundant quantities of water [72].

*Rabdosia rubescens* has two extracts from this herb, oridonin and ponicidin, which have been shown in the laboratory to possess anticancer properties, but more research is needed to determine exact effects as well as a suggested dosage [73].

*Saw palmetto*, found by Native Americans, is commonly used to treat symptoms of benign prostatic hyperplasia, or an enlarged prostate, such as difficult or frequent urination. Saw palmetto may also be used in an effort to prevent or treat prostate cancer, although the results of such practice are unclear [74, 75].

*Sea vegetables*, primarily be noted as algae and seaweed, have been shown to have anti-cancer properties as eating more sea vegetables, opposed to red meat, may result in a lower risk of breast cancer. In addition some laboratory research has shown that algae have the potential to slow the cancer cell growth rate and kill cancer cells [76].

*Shark liver oil* contains chemicals that may have anti-cancer properties, such as alkylglycerols, squalene, and squalamine, which have both shown promise in laboratory tests [77].

*Shiitake mushrooms* are an edible fungus that grows in the forests of Asia and have been shown to have cholesterol lowering, anti-viral, and anti-cancer properties in laboratory studies [78].

*Soy*, soybeans, soy oil, and/or other soy food products have been shown to lower blood pressure and cholesterol, while also having the potential to help prevent certain cancer types such as that which affects the breasts or prostate [79, 80, 81].

*St. John's wort* has been shown to work as an antidepressant with the power to treat mild to moderate depression and anxiety, although it can have harmful interaction effects if used with other drugs of any kind [82, 83].

*Tea tree oil*, also know as melaleuca oil, comes from the Australian tea tree and is used externally to cleanse open sores or wounds, improve the overall condition of the skin, and help the epidermis to heal more quickly [84].

*Turmeric* is a root that has been used for hundreds of years in Chinese and Ancient Indian medicine, as the active ingredient, curcumin, is an antioxidant that has been shown in the laboratory to attack free radical activated oxygen molecules, as it interferes with the development, growth, and spread of cancer cells. Turmeric may also be used to help to reduce inflammation, improve the functioning of the liver, aid digestion, regulate menstruation, and relieve arthritis pain [85, 86].

*Valerian root* has been most commonly used for many generations in teas or in pill form as a sleep aid, although it may also be effective to help treat anxiety [87, 88].

*Vitamin C*, also noted as ascorbic acid, is an antioxidant that comes from fruits and vegetables, and cannot be made by the human body. Studies have noted that a diet high in vitamin C helps prevent disease, significantly reduces the risk against many types of cancer, and may help to extend the life of cancer victims [89].

*Vitamin D* is important to the body as it is needed to regulate the amount of calcium and phosphorus in the body using the calcium to build bones to help keep them strong. Vitamin D is made by the body after exposure to the ultra violet rays of the sun and is also available through certain foods or supplements. Some laboratory studies have indicated that higher intake of vitamin D may be linked to a lower risk of cancer, although overexposure to sunlight can increase the risk of developing skin cancer [90].

# Complementary Therapies

There are many various types of complementary therapies that may be able to help anybody to live a happier, more productive, less stress filled life. Some of these complementary treatments may include: apitherapy, aromatherapy, art therapy, Asian bodywork, Ayurvedic medicine, biofeedback, bodywork, breathwork, child alternative therapies warning, Chinese herbal medicine, chiropractic treatment, dance therapy, electrodermal screening, faith healing, feng shui, humor therapy, hydrotherapy, Kampo, massage, music therapy, native American healing, physical activity, polarity therapy, psychotherapy, qigong, reflexology, reiki, shamanism, spirituality and/or tai chi. Although, a cancer patient should consult their doctor before beginning to using complementary treatment therapy to insure there are no treatment interactions.

*Apitherapy* (aka. bee venom, bee venom therapy, bee pollen, bee venom immunotherapy) is noted by the use of various common honey bee products, such as raw honey, pollen, royal jelly, venom propolis (substance bees create to coat inside of hives), and melittin. This treatment is thousands of years old and may be useful as practitioners have noted that bee venom's anti-inflammatory

22

properties can be utilized to treat lower back pain, migraine headaches, and chronic pain. Other proponents have noted that bee pollen has five to seven times more protein than beef, that honey comb containing bee pollen has effectively treated allergies, and bee pollen has also been noted to increase endurance, energy, and overall performance. Others have made the claim that raw honey is an energy building source that contains B complex vitamins and may have antibacterial, anti-inflammatory, anti-fungal, as well as anti-tumor properties. Bee products have been used to treat other unnoted conditions in addition to some of the symptoms of cancers and side effects of treatment. During classic bee venom treatment a practitioner will generally use live bees to sting the patient at specific trigger points to reduce pain and inflammation. Other bee products, such as honey and pollen, are generally taken orally or via injection. Bee products can generally be found at health care stores, pharmacies, on the internet, and in stores that specialize in such products. Although, none of these products are FDA approved and individuals should consult their doctor before beginning to take such products on a regular basis [91].

*Aromatherapy* is thousands of years old as it consists of the use of essential oils and fragrant substances that when inhaled may improve health or alter mood, as well as support the balance of mind, body, and spirit. This therapy is simply a treatment for symptoms of cancer or side effects of cancer treatment generally in combination with other therapy. When highly concentrated aromatic substances are distilled from herbs such as lavender, lemon, eucalyptus, rosemary, jasmine, peppermint, chamomile, geranium, majoram, bergamot, cederwood, and tea tree, among other substances, are generally applied to the skin (or possibly in an aromatic bath) and then inhaled like a "vaporub", since they have been shown to improve symptoms such as stress, nausea, anxiety, pain, high blood pressure, irregular respiratory rates, tension, insomnia, and depression for centuries. It is commonly believed that olfactory nerve receptors in the nasal cavity send chemical messages through nervous pathways to the limbic system in the brain in order to affect emotions, moods, respiration, heart rate, and blood pressure. Others say that oils may be absorbed through the skin. I believe it is much simpler than that since the smoke from the burning herbs or aroma from strong concentrated oils is inhaled it works in the same way that oxygen and inhaled smoke from drug use work as they are both absorbed by the lungs, then by blood cells which deliver the drugs throughout the body. The way that the NCI describes, aromatherapy works in ways similar to cocaine, as it is absorbed in the sinus cavity and lungs while the rest drains down the esophageal track into the stomach. My way is much simpler and seems obvious since aromatherapy is advantageous to the cardiovascular and respiratory systems. These therapies are generally noted as safe, but beware of allergic reactions, as well as hormone like effects of lavender and tea tree oil which may block or decrease the effects of androgens (male sex hormones) [92, 93].

*Art therapy* is a creative and expressive therapy in which an individual or group of individuals express themselves through artwork such as painting, drawing, or sculpting. During this therapy individuals are often able to set aside their feelings of physical pain and get in better touch with their conscious and unconscious emotions and feelings. As a result, these people tend to be better able to deal with their cancer symptoms and side effects of treatment. This therapy is especially beneficial for children and adults who often are unable to express themselves well verbally, as this therapy helps individuals to increase their self awareness and come to terms with their emotional conflicts through artistic expression. Art therapy is generally safe, but people should understand how to use tools for creating their artwork to do so responsibly. The act of creating your own artistic masterpiece is usually a very uplifting experience [94].

*Asian Bodywork* is a medicine that is based upon the principles of qi ("ch'i"), the life force of vital energy which flows through twelve meridians or channels in the human body. This theory notes that qi is present in all living things, as meridians direct the flow of energy throughout the body internally. If energy flow is out of balance, or blocked, illness may occur. These medicinal practices are intended to help restore balance within the body and tend to be beneficial to the body and psyche by stimulating acupoints noted along the meridians of the body. Individuals with a terminal condition as well as cancer should talk to a doctor before such treatment. Asian bodywork may include acupuncture, acupressure, moxibustion, ohashiatsu, shiatsu, tui na and/or watsu.

Acupuncture is an ancient Chinese practice that may be used alone or in combination with other treatments to take care of side effects caused by cancer therapy and symptoms of cancer, but this treatment does not affect cancer cells by itself. This treatment consists of a trained individual using sterile, one time use, acupuncture needles which are inserted into specific points around the body in order to balance the body's energy, or qi (ch'i) as noted in China. This method is thousands of years old and is used to treat pain, nausea, addictive behavior, and is sometimes used as an anesthetic. Variations include acupressure, sonopuncture, and electroacupuncture; as other variations may include the use of suction, friction, heat, lasers, or magnets directed at acupoints.

Acupressure is basically acupuncture without needles, using fingers, other body parts, or devices to apply pressure to acupoints as well as massage, stretching, and other specific methods noted. Note: Methods may increase the chance of metastases in a tumor, bone, and tissue cancer as deep pressure to such areas and surrounding tisss may encourage cancer cells to spread.

Moxibustion, which is commonly practiced in China and Tibet, consists of the use of smudge bundles consisting of specific herbs which when burned and then the heat from the bundle is applied to acupoints in order to help to open up and stimulate them helping the body become better able to heal itself.

Ohashiatsu is a method based on shiatsu, but is focused more toward restoring an energy balance to the entire body rather than just a part of the body. The success of this method depends upon the technical skill of the practitioner as well as the feelings of empathy and compassion conveyed by the practitioner to the client. The ultimate goal is inner peace and harmony. This method also uses meditation and exercise.

Shiatsu (Japanese word which means "finger pressure") consists of the application of pressure on energy meridians and acupoints to stretch out and open qi pathways in the body, as subtypes of this method focus on breathing techniques, stretching, meditation, and other practices.

"Tui na" may predate acupuncture as this method uses many various massage techniques to apply pressure to acupoints in an effort to open meridians. The spinal muscles are often the target of this treatment.

Watsu (aquatic shiatsu) is practiced in warm water as the practitioner holds the individual afloat while the patient is massaged, stretched, and cradled by the practitioner in an effort to help release the body's physical and emotional stress by opening the meridians of

28

the body. A level of connection and trust may develop between patient and practitioner as the patient is held afloat during treatment. People with bowel control problems, infections, or fevers should not use watsu [95, 96, 97, 98].

*Ayurvedic medicine* is an ancient treatment from India predates written text, having been handed down by word of mouth by previous generations. Ayurvedic medicine practiced in India is noted in two ancient texts written in Sanskrit, known as the "Caraka Samhita" and the "Sushruta Samhita." Ayurvedic medicine at the most basic level can be noted by how herbal, plant, spice, and/or mineral remedies aim to restore balance and integrate the mind, body, and spirit. In India over 150 undergraduate colleges and 30 graduate colleges primarily teach the practices of Ayurvedic medicine. Ayurvedic medicine is based upon the interconnectedness of the universe and everything contained within it, as humans as other creatures and living things are composed of elements that are found throughout the universe. According to this philosophy it is noted that when an individual's mind, body, and spirit are in harmony with the universe as a result of natural and wholesome interactions with it a person can expect good health. While the individual who is out of harmony with the universe as a result of spiritual, emotional, physical, or a combination of such problems can expect disease to arise. Ayurvedic medicine also takes note of an individual's "constitution" or prakriti, which is a person's general physical condition, psychological disposition, and the way in which the body maintains homeostasis. The prakriti of a person is believed to remain unchanged throughout their life and is characterized by three doshas (life forces) which control the mind and body's activity. The doshas each are composed of 2 of the 5 basic elements; earth, water, fire, air, and ether (upper regions of space). The three doshas are noted as the kapha dosha, pitta dosha, and vata dosha [99, 100].

*Biofeedback* is a modern meditative practice in which an individual focuses on and hopefully modifies a bodily function or process. Usually the therapist begins with a simple function such as breathing rate and heart rate. Bodily functions are monitored as prompted meditation begins with patient and practitioner by finger impulse, thermal biofeedback, electrodermal activity, electromyogram, as well as monitoring breathing rate. Biofeedback helps to improve quality of life, not likely to work for cancer [101].

*Bodywork* is a complementary technique in which a practitioner helps the individual to correct posture, realign the body, manipulate soft tissue or joints, and may help individual to gain more self awareness of own body. Three bodywork techniques noted: rolfing, Alexander technique, Trager approach, and Feldenkrais method [102].

*Breathwork* is a relaxation technique noted by focusing on exaggerating the ways in which a person naturally inhales and exhales. Long Deep Breaths in almost a meditative manner may help to reduce stress, anxiety, and tension over time. There are many different types of breathwork and facilitators that help individuals and/or groups during therapy. The person usually lies on the floor for this therapy during relaxation [103].

*Child alternative therapies* WARNING!!! Children do not react to alternative medicine or complementary therapy like adults. Beware interactions with other drugs and treatments. Consult ALL DOCTORS FIRST!!! Although, support groups, family, and child group/individual counseling therapy can be very beneficial for a child to gain a better understanding and acceptance of the situation at hand. Humor and play therapies can also be very beneficial for a child's psychological well being and to keep the young one optimistic about other treatment [104].

*Chinese herbal medicine,* as a part of traditional Chinese medicine these methods of cancer symptom treatment have been used for more than two thousand years. This treatment consists of more than 500 herb extracts or single herbs and almost 300 complex formulas, using more than 3,000 herbs and 300 mineral/animal extracts. The formulas commonly contain between four and twelve ingredients, with one or two having the greatest effect and the others being in place to treat other minor aspects of the problem being treated. Always consult a practitioner of Chinese medicine before beginning symptom treatment. Although in the United States herbs can be purchased at some pharmacies, healthcare stores, as well as from herbal practitioners [105].

*Chiropractic treatment* is commonly used by a trained professional that treats body pain in the back, as well as other parts of the body by means of physical bone, joint, and muscle manipulation [106, 107].

*Dance therapy* is a movement therapy that can be physically, psychosocially, and psychologically beneficial. This therapy is based on the idea that body and mind are both used in conjunction during dance. Emotional responses are often evident during dance. Be sure bone, joint, and cardiovascular issues will not be problematic. Ideally a practitioner will suggest specific dance moves that may be beneficial to learn and improve your condition. Usually the cancer victim and possibly caregivers dance in a group setting after the initial interview [108].

*Electrodermal screening* is electroacupuncture that is used as a diagnostic tool to determine if energy imbalances exist, using electrical resistance of skin's surface to measure. A chiropractor, acupuncturist, naturopath, homeopath, or other practitioner may use this method, which is based on galvanic skin tests which originated in the early twentieth century [109].

*Faith healing* involves belief in a higher power and possibly advice from spiritual leader(s) as this positive belief and good faith may be beneficial complementary therapy during treatment. Faith in God and prayer can reduce stress and anxiety, promote peace of mind, and increase a person's will to live. Although, faith healing as a primary treatment is noted as behavior that is very risky to health and may result in cancer symptoms speeding up since medical therapy is not being used [110].

*Feng shui* is a practice that is based on the on the Chinese belief that everything has a balance to it. The art of feng shui is based on the placement of everything around you in order to find perfect harmony. When feng shui is found by a person they feel closer in harmony with the universe and everything within it [111].

*Humor therapy* utilizes humor to essentially help individuals to gain a better personal disposition, improve their quality of life, as well as deal with emotional stress and physical pain. Humor therapy can also have positive effects on the immune system, the circulatory system, and other systems of the body [112].

*Hydrotherapy* involves the use of $H_2O$ in one of its states of matter as solid ice, liquid water, or gaseous steam, as a medical treatment of the symptoms of cancer internally or externally. Hydrotherapy may be used to alleviate small aches and pains with, for physical therapy, and to promote relaxation. Water can be used to drink in order to prevent dehydration; clean wounds; in ice packs to reduce swelling and slow circulation; in warm wraps, steam baths, saunas, humidifiers, and hot tubs to increase circulation and relax

34

muscles; in a pool as a passive physical therapy to help to strengthen muscles by performing exercises with a reduced amount of strain on bones and joints than regular physical therapy while offering resistance to movement effectively building muscle. Of course water helps us all to stay clean, as warm water and cleansing baths both are important for the body to relax and stay free from disease [113]. Hydrotherapy and Aromatherapy can also be combined in a hot, fragrant, bath soak as well!

*Kampo* is a Japanese herbal medicine that consists of more than 210 herbal remedies, which are prepared in set combinations, may be useful to help treat the symptomatic side effects of cancer and cancer treatment. This treatment is based on the idea of a balance to the human body and all of the different systems within it. A Kampo practitioner tends to prescribe remedies in order to treat specific symptoms that the patient is suffering from in order to bring balance back to the body [114].

*Massage* has been used for many millennia as it helps to relax the patient by physically kneading, rubbing, and manipulating the soft tissues and muscle of the body. This type of therapy may help to relieve both physical and mental tension, although it may also promote metastasis in tissue cancers, so patients with such cancers should use this relaxation technique with caution [115, 116].

*Music therapy* is used to help individuals, as well as groups of people, to improve their personal quality of life by listening to, creating, or participating in the creation of music as a form of personal expression that can have positive physical and mental effects as a result [117].

*Native American healing* is based on the 40,000 year old Native American belief in the interconnectedness of everything in the universe, on the planet, and wherever we are. Healers of sort; such as "medicine men", herbalists, shamans, and spiritual leaders; may use various purifying processes, herbal remedies, shamanism, and/or spiritual healing rituals in order to treat the symptoms of cancer and cancer treatment side effects. Native American healing has been shown to help improve the quality of life by reducing the pain and stress through prayer, introspection, spiritual support, community support, and meditation [118].

*Physical activity* involves regular exercise, even as little as a daily walk or bike ride, during cancer treatment can help to improve stamina, control weight, help better blood flow, improve personal mental and physical disposition, as well as the patient's overall quality of life. Of course every individual's personal exercise regime may be different based on their physical ability, as well as other factors, but should stay fun [119].

*Polarity therapy* consists of a practitioner identifying sources of energy blockage throughout the body by noting areas of tension, pain, and discomfort. Relaxation and improved disposition is promoted by increasing range of motion and overall energy, while relieving pain and tension, in addition to reducing swelling and inflammation using physical bodily manipulation in combination with massage, breathing techniques, diet changes, physical therapy techniques, and supportive counseling [120].

*Psychotherapy* is defined as therapy of the mind which may focus on personal problems and/or personal disposition as it may be beneficial to an individual's personal life outlook, perception of the world, and interpersonal interactions. The best type of this therapy that does not cost anything is to talk out your problems with a person or people that you trust which will likely help you begin to feel better. People who can for you always have an open ear and are usually ready to listen. There are many different types of psychotherapy techniques that may be used to help patients to deal with their problems like depression, anxiety, relaxation, coping with changes due to cancer, and improvement of quality of life.

Behavioral therapy is utilized to effectively modify disruptive behavior and improve human functioning based on classical and operant conditioning. Classical conditioning can be noted when a neutral stimulus is able to cause the response that was initially produced by another stimulus. An example of classical conditioning may noted in an experiment when at the beginning a puff of air is blown in an individual's eye to make them blink at the same time a bell rings. Then at the end of the experiment the individual blinks just at the sound of a bell since the stimuli came at the same as the puff of air so many times before. The behavioral therapy more commonly used with cancer patients is operant conditioning. Operant conditioning can be noted as rewards for beneficial behaviors to help increase those behaviors and punishment for behaviors that are not beneficial. As a result there is an increased likelihood that more beneficial and fewer negative behaviors are performed.

Cognitive therapy is utilized to help modify thoughts and though processes in order to help improve emotions, perceptions, and behavior.

Personal and/or family counseling may be utilized to help resolve intrapersonal and interpersonal conflicts by helping to improve understanding of oneself and others around you.

Hypnosis is sometimes utilized, as a hypnotist or therapist will use various forms of suggestion to help an individual to experience changes in cognition, perception, sensation, or control over motor behaviors.

Meditation is utilized to help relaxation, focus energy, to resolve personal problems of various types, and may help cancer treatment through a conscientious control of subconscious bodily functions, such as the immune system, which is mind-body medicine. Meditation is most commonly performed in a relaxed position and generally in a quiet or very comfortable environment where an individual or group focuses their personal energy to resolve physical, mental, and/or interpersonal problems.

Support groups are a great source of hope, understanding, knowledge, and support while dealing with the symptoms of cancer and side effects of cancer treatment. There are many support groups that are based on cancer type, some of which are noted at the end of this chapter, as individuals with the same cancer can empathize with each other regarding treatment, effects, difficulties, and progression of the cancer. Other support groups may based on cancer staging, educational intervention, coping skills, general mental health, for children of cancer victims, and youth victims of cancer. Support groups for cancer victims may be available in person, online, and over the telephone, while some of them may have open membership and others have closed membership. Support groups are generally beneficial to those people who take part in them as they are able to share their emotions, find interpersonal support, find tips about dealing with symptoms, and feel less isolated as a victim of cancer or any other negative condition.

*Qigong*, also noted as spiritual qigong, has been practiced by the Chinese for thousands of years and is used to help improve the body's qi energy flow and enhance the quality of life, as well as self awareness through the use of breathing exercises, meditation, and exercising specific areas of the body [121].

*Reflexology*, which has been used since the times of ancient Egypt and China, utilizes pressure on specific areas of muscular groups in an effort to relieve tension, promote relaxation, and reduce pain. This therapy may relieve some of the physical symptoms of cancer for a relatively brief period of time, likely a few days at the most [122].

*Reiki*, which has been practiced by the Japanese for thousands of years, is based on the concept of "universal life energy", which is the term from Japan defined. As a practitioner works to help heal the cancer victim's symptoms by promoting relaxation and the sense of well being using the energy flow through the hands placed above the effected area for various amounts of time in an effort to balance the patient's energy flow. Reiki is a spiritual therapy that has been shown to improve a cancer victims' personal disposition, as well as reduce nausea, vomiting, and pain [123, 124].

*Shamanism* is an ancient practice, which is over 40,000 years old from East Asia has utilized various natural organic substances, in combination with the use of imagery and spirituality to help treat the symptoms of cancer and side effects of cancer treatment. At the least this therapy helps individuals to reduce anxiety, stress, as well as the physical symptoms of cancer and treatment. At best this therapy has the potential to help cancer to begin to recede and shrink, but that is not guaranteed and may be a result of a more optimistic and energized patient rather than the herbal therapy itself [125].

*Spirituality* can be noted throughout the world in many different specific forms of religion and other specific beliefs. A cancer patient's quality of life can be enhanced by prayer and interactions with spiritual leaders, hopefully gaining a greater will to live, a more optimistic outlook, and helping patients to become better able to deal with their life problems. While science does not support spiritual healing by itself, prayer and religious support have shown to have positive effects on the healing and recovery processes. An individual's inner peace, interconnectedness with others and the universe, personal beliefs, and personal meaning of life may be expressed through spirituality, as faith and spirituality is important for recovery from cancer [126, 127].

*Tai Chi* is an ancient Chinese martial art division of qigong that involves mind-body practice. The physical movement exercises and breathing techniques used during Tai Chi can help to better cardiovascular circulation and help overall relaxation. Tai chi as part

42

of qigong is intended to help restore qi balance, as this technique, when regularly practiced, has been proven to improve physical bodily attributes and increase stamina [128, 129].

# Symptoms and Treatments

The pains of daily living can be stressful and hard on us all in many various ways. The physical stressors of work, family life, and recreational activities can be difficult to deal with every day as they sometimes spill over into family and social life, as well as slowing the progress of work. When we blow up on a significant other as a result of too much on the mind from various life stressors, it is then obvious that stress relief is needed. There are many herbal supplements and complementary therapies that can help to relieve the mental stress, physical, and other life problems.

There are supplements for unhealthy bowels, lower cholesterol, muscle pain, joint pain, swelling/inflammation, diarrhea, a laxative, circulatory problems, eye irritation, skin conditions, ulcers, frequent urination, appetite stimulus, an enlarged prostate, premenstrual syndrome, rheumatoid arthritis, stomach ache, tooth ache, respiratory problems, pressure behind eyes, heart disease, nausea, oral pain, back pain, improve heart rate, high blood pressure, mouth sores from cancer treatment, menstrual cramps, menopausal symptoms, night sweats, hot flashes, heal external injuries, to help prevent cancer, help fight cancer, help fight and prevent viruses, improve energy, help chronic conditions, sooth skin irritation, help with weight loss, help improve body's homeostasis, regulate menstruation, help prevent disease, improve immune system, reduce fatigue, reduce effects of cancer and treatment, reduce chemo side effects, prevent bone loss or fragile bones, improve mood, improve sleep quality, improve immune system, improve circulation, improve thought processes, improved mental alertness, improve liver functions, and improve digestion.

There is complementary therapy for back pain, fatigue, swelling/inflammation, migraine headaches, chronic pain, allergies, nausea, physical stress relief, high blood pressure, tension, respiratory problems, insomnia, fungal growth, bacteria problems, to improve bodily functions/processes, control weight, improve immune system function, slow circulation, improve circulation, improve stamina, reduce swelling, alleviate aches and pains, to help build muscle, physically feel better, improve body's energy flow, gain harmony with world and universe around you, to improve social life, psychologically feel better, beat depression, improve mood, relieve anxiety, increase will to live, better deal with emotional stress, relive mental tension, improve general temperament, obtain peace of mind, improve quality of life, gain greater self expression, increase self awareness, gain better personal disposition, improve personal life outlook, come to terms with emotional conflict, improve body's personal balance; as well as helping to improve mind, body, and spirit.


Addictive Behavior - *Asian Bodywork,* **Canarius,** *Psychotherapy, Spirituality*

Allergies - *Apitherapy,* **Canarius**

Anxiety - *Aromatherapy, Breathwork,* **Canarius,** *Faith healing, Kava, Lavender, Marijuana, Music therapy, Physical activity, Psychotherapy, Shamanism, Spirituality, St. John's wort, Valerian root*

Appetite Poor – **Canarius (after therapy is complete),** *Marijuana, Physical activity, Psychotherapy*

Arthritis - **Canarius,** *Chinese thunder god vine, Comfrey, Flaxseed, Turmeric*

Back Pain - *Apitherapy, Bodywork,* **Canarius,** *Chiropractic treatment, Comfrey, Feverfew, Hydrotherapy, Massage, Reflexology*

Bacteria Problems - *Antioxidants, Apitherapy,* **Canarius**

46

Bodily Functions (Personal Balance/Homeostasis Improvement) - *Antioxidants, Ayurvedic medicine,* **Canarius,** *Physical activity, Psychotherapy, Reiki*

Bone Fatigue - *Calcium,* **Canarius,** *Vitamin D*

Bowels Unhealthy - *Acidophilus,* **Canarius**

Cancer and Medical Treatment Side Effects - *Astralagus, Ayurvedic medicine,* **Canarius,** *Chinese herbal medicine, Feng shui, Ginger, Kampo, Native American healing, Psychotherapy, Reiki, Shamanism, Spirituality*

Cancer Fighting - *Astralagus,* **Canarius,** *Chinese thunder god vine, Chlorella, Clove supplements, Copper, Ellagic acid, Flaxseed, Garlic, Ginseng, Grapes/seed extract, Green tea, Indian snakeroot, Licorice root extract, Lycopene, Maitake mushrooms, Melatonin, Milk thistle, Modified citrus pectin, Noni plant, Omega-3 fatty acids, Pokeweed, Rabdosia Rubescens, Saw palmetto, Sea vegetables, Shark liver oil, Shiitake mushrooms, Turmeric, Vitamin C, Apitherapy*

Cancer Prevention - *Acidophilus, Antioxidants, Calcium,* **Canarius,** *Chinese thunder god vine, Chlorella, Folate, Garlic, Ginseng, Grapes/seed extract, Green tea, Licorice root extract, Lycopene, Maitake mushrooms, Melatonin, Noni plant, Saw palmetto,*

*Sea vegetables, Shiitake mushrooms, Soy, Turmeric, Vitamin C, Vitamin D*

Cardiovascular (Circulatory Problems) - *Aromatherapy, Bilberry,* **Canarius,** *Gingko, Gotu kola, Grapes/seed extract, Green tea, Hawthorn, Horse chestnut, Omega-3 fatty acids, Humor therapy, Hydrotherapy, Physical activity, Tai Chi*

Chemotherapy Side Effects - *Bromelain,* **Canarius**

Chronic Conditions - **Canarius,** *Noni plant*

Chronic Pain - *Apitherapy,* **Canarius**

Cognition/Thought Processes improvement - **Canarius,** *Ginseng, Physical activity, Psychotherapy*

Constipation - *Aloe Vera plant oil,* **Canarius,** *Castor oil, Flaxseed oil, Physical activity, Psyllium*

Depression - *Aromatherapy,* **Canarius,** *Lavender, St. John's wort, Physical activity, Psychotherapy, Spirituality*

Diarrhea - *Bilberry,* **Canarius**

Digestion Problems - **Canarius,** *Green tea, Peppermint oil, Turmeric*

Disease Prevention - *Astralagus,* **Canarius,** *Vitamin C*

Emotional Conflict - *Art therapy,* **Canarius,** *Psychotherapy, Spirituality, Tai Chi*

Emotional Stress - *Aromatherapy, Art therapy, Asian Bodywork,* **Canarius,** *Dance therapy, Humor therapy, Psychotherapy, Spirituality, Tai Chi*

Energy Flow Improvement - *Apitherapy, Asian Bodywork, Astralagus,* **Canarius,** *Chiropractic treatment, Electrodermal screening, Feng shui, Native American healing, Physical activity, Polarity therapy, Psychotherapy, Qigong, Reflexology, Reiki, Tai Chi*

Eye Irritation - **Canarius,** *Castor oil*

Eye Pressure - **Canarius,** *Marijuana*

Fatigue - *Astralagus,* **Canarius,** *Carnitine, Ginseng, Green tea, Apitherapy, Physical activity*

Frequent Urination - **Canarius (after therapy is completed),** *Saw palmetto*

Fungal Growth - *Apitherapy,* **Canarius**

Hair Loss - *Lavender*

Harmony With World and Universe Around You - *Ayurvedic medicine,* **Canarius,** *Chinese herbal medicine, Feng shui, Kampo,*

*Native American healing, Physical activity, Psychotherapy, Qigong, Reiki, Spirituality, Tai Chi*

Headache - **Canarius,** *Feverfew, Lavender*

Heart Disease - **Canarius,** *Garlic, Omega-3 fatty acids*

Heart Rate Problems - **Canarius,** *Copper, Hawthorn*

High Blood Pressure - *Aromatherapy,* **Canarius,** *Copper, Garlic, Grapes/seed extract, Indian snakeroot, Indian snakeroot, Omega-3 fatty acids, Soy*

High Cholesterol – *Acidophilus,* **Canarius,** *Flaxseed, Garlic, Grapes/seed extract, Green Tea, Milk thistle, Omega-3 fatty acids, Psyllium, Shiitake mushrooms, Soy*

Hot Flashes - *Black cohosh,* **Canarius**

Immune System Problems - *Astralagus root,* **Canarius,** *Clove supplements, Grapes or seed extract, Humor therapy, Maitake mushrooms, Pokeweed, Vitamin C*

Immune System Overactive - **Canarius,** *Chinese thunder god vine*

Infection - *Arnica,* **Canarius,** *Tea Tree Oil*

Insomnia - *Aromatherapy,* **Canarius (after therapy is completed),** *Carnitine, Gotu kola, Grapes/seed extract, Indian*

50

*snakeroot, Kava, Lavender, Marijuana, Melatonin, Music therapy, Psychotherapy, Spirituality, Valerian root,*

Joint Pain - *Arnica, Bodywork,* **Canarius,** *Chiropractic treatment, Hydrotherapy, Noni plant, Polarity therapy*

Life Quality Problems - *Ayurvedic medicine, Biofeedback,* **Canarius,** *Humor therapy, Music therapy, Psychotherapy, Qigong, Spirituality, Tai Chi*

Liver Function Problems - **Canarius,** *Milk thistle, Turmeric*

Memory Problems - **Canarius,** *Gotu kola, Psychotherapy*

Menopausal Symptoms - *Black cohosh,* **Canarius,** *Chasteberry*

Menstrual Cramps - *Bilberry,* **Canarius**

Menstruation Regulation - **Canarius,** *Chinese thunder god vine, Flaxseed, Turmeric*

Mental Alertness Problems - *Apitherapy,* **Canarius,** *Green Tea, Physical activity, Psychotherapy, Qigong*

Mental Tension - **Canarius,** *Dance therapy, Massage, Music therapy, Psychotherapy, Qigong, Spirituality, Tai Chi*

Migraine Headache - *Apitherapy,* **Canarius,** *Psychotherapy*

Mind, Body, and Spirit Balance - *Aromatherapy, Asian Bodywork, Ayurvedic medicine, **Canarius**, Feng shui, Psychotherapy, Qigong, Spirituality, Tai Chi*

Mood Improvement - ***Canarius,** Carnitine, Native American healing, Psychotherapy, Qigong, Spirituality, Tai Chi*

Mouth Sores From Cancer Treatment - ***Canarius,** Capsicum*

Muscle Building - *Bodywork, **Canarius,** Hydrotherapy*

Muscle Pain - *Arnica, Asian Bodywork, Bodywork, **Canarius,** Comfrey, Chiropractic treatment, Feverfew, Hydrotherapy, Massage, Polarity Therapy, Reflexology, Reiki,Tai Chi*

Nausea - *Aromatherapy, Asian Bodywork, **Canarius,** Ginger, Lavender, Marijuana, Peppermint oil, Reiki*

Night Sweats - *Black cohosh, **Canarius***

Pain - *Apitherapy, Arnica, Aromatherapy, Asian Bodywork, Bodywork, **Canarius,** Capsicum, Clove supplements, Comfrey, Chiropractic treatment, Feverfew, Kava, Humor therapy, Hydrotherapy, Marijuana, Massage, Native American healing, Noni Plant, Polarity therapy, Reflexology, Reiki, Tai Chi, Turmeric*

Peace of Mind - *Art therapy, **Canarius**, Faith healing, Psychotherapy, Spirituality, Tai Chi*

Personal Disposition Improvement - *Art therapy, Asian Bodywork, Ayurvedic medicine,* **Canarius,** *Chinese herbal medicine, Feng shui, Humor therapy, Music therapy, Kampo, Native American healing, Physical activity, Psychotherapy, Qigong, Reiki, Spirituality, Tai Chi*

Personal Life Outlook Improvement - *Art therapy, Asian Bodywork,* **Canarius,** *Humor therapy, Physical activity, Psychotherapy, Spirituality, Tai Chi*

Premenstrual Syndrome - **Canarius,** *Chasteberry*

Prostate Enlarged - *Saw Palmetto*

Relax - *Asian Bodywork, Biofeedback, Bodywork, Breathwork* **Canarius,** *Chiropractic treatment, Dance therapy, Kava, Massage, Humor therapy, Hydrotherapy, Music therapy, Polarity therapy, Physical activity, Psychotherapy, Qigong, Reflexology, Spirituality, Tai Chi*

Respiratory Problems - *Aromatherapy, Breathwork ,* **Canarius,** *Flaxseed, Licorice root extract, Peppermint tea, Tai Chi*

Self Awareness Improvement - *Art therapy, Bodywork,* **Canarius,** *Polarity therapy, Psychotherapy, Qigong*

Self Expression Improvement - *Art therapy,* **Canarius,** *Music therapy, Psychotherapy*

Skin Conditions - *Aloe, Arnica,* **Canarius,** *Castor oil, Noni plant, Tea tree oil*

Skin Irritation - Arnica, **Canarius,** *Tea tree oil*

Sleep Problems - *Aromatherapy,* **Canarius (after therapy completed),** *Carnitine, Gotu kola, Grapes/seed extract, Indian snakeroot, Kava, Marijuana, Melatonin, Music therapy, Psychotherapy, Valerian root*

Social Life Problems - **Canarius,** *Dance therapy, Music Therapy, Psychotherapy, Physical activity, Spirituality*

Stamina Improvement - *Apitherapy,* **Canarius (after therapy completed),** *Physical activity, Tai Chi*

Stomach Ache - **Canarius,** *Feverfew, Shamanism*

Stress - *Aromatherapy, Bodywork, Breathwork,* **Canarius,** *Faith healing, Gotu kola, Indian snakeroot, Kava, Music Therapy, Hydrotherapy, Massage, Native American healing, Polarity therapy, Psychotherapy, Qigong, Reflexology, Shamanism, Spirituality*

Swelling/Inflammation - *Apitherapy, Arnica, Bromelain,* **Canarius,** *Chinese thunder god vine, Comfrey, Gotu kola, Polarity therapy, Tai Chi, Turmeric*

Tension - *Aromatherapy, Bodywork, Breathwork,* **Canarius,** *Gotu kola, Indian snakeroot, Kava, Hydrotherapy, Massage, Music Therapy, Native American Healing, Polarity therapy, Psychotherapy, Qigong, Reflexology, Shamanism*

Thought Processes/Cognition Improvement - **Canarius,** *Ginseng, Music Therapy, Psychotherapy, Qigong, Spirituality*

Tooth Ache - **Canarius,** *Clove supplements, Feverfew*

Ulcers - **Canarius,** *Licorice root extract*

Virus Preventing and Fighting Agents - *Antioxidants,* **Canarius,** *Licorice root extract, Pokeweed, Shiitake mushrooms*

Vomiting - **Canarius,** *Ginger, Marijuana, Reiki*

Weight Control - **Canarius,** *Green tea, Physical activity*

Will To Live Improvement - *Art therapy,* **Canarius,** *Faith healing, Humor therapy, Native American Healing, Psychotherapy, Spirituality*

Wound Healing – *Arnica,* **Canarius,** *Comfrey, Hydrotherapy, Tea tree oil*

# Conclusion

With all the pains that life can bring there are always ways to make living easier and more enjoyable under any circumstances. All negative conditions and ailments of the human body, spirit, and mind can be remedied to some degree. People have recovered from and beaten even the worst of things in which there seemed to be no way to stop the torment of their agony. Positive thinking and knowledge of solutions to treat all of these horrible negative human conditions in combination with initiative to actually apply them is all it takes to get better! Other natural cancer remedies are noted at whale.to/cancer/therapies.html

# References

1.  American Cancer Society. Acidophilus. American Cancer Society. <http://www.cancer.org/Treatment/TreatmentsandSideEffects/Complementar yandAlternativeMedicine/DietandNutrition/acidophilus> (Accessed August 2010).

2.  American Cancer Society. Aloe. American Cancer Society. <http://www.cancer.org/Treatment/TreatmentsandSideEffects/Complementar yandAlternativeMedicine/HerbsVitaminsandMinerals/aloe> (Accessed August 2010).

3.  National Center for Complementary and Alternative Medicine. Aloe Vera. National Center for Complementary and Alternative Medicine. <http://nccam.nih.gov/health/aloevera/> (Updated June 2010). (Accessed August 2010).

4.  National Cancer Institute. National Cancer Institute FactSheet – Antioxidants and Cancer Prevention: Fact Sheet. National Cancer Institute. <http://www.cancer.gov/newscenter/pressreleases/antioxidants> (Updated July 28, 2004). (Accessed August 2010).

5.  American Cancer Society. Arnica. American Cancer Society. <http://www.cancer.org/Treatment/TreatmentsandSideEffects/Complementar yandAlternativeMedicine/HerbsVitaminsandMinerals/arnica> (Accessed August 2010).

6.  American Cancer Society. Astragalus. American Cancer Society. <http://www.cancer.org/Treatment/TreatmentsandSideEffects/Complementar yandAlternativeMedicine/HerbsVitaminsandMinerals/astragalus> (Accessed August 2010).

7.  National Center for Complementary and Alternative Medicine. Astragalus. National Center for Complementary and Alternative Medicine. <http://nccam.nih.gov/health/astragalus/> (Updated July 2010). (Accessed August 2010).

8.  National Center for Complementary and Alternative Medicine. Bilberry. National Center for Complementary and Alternative Medicine.

&lt;http://nccam.nih.gov/health/bilberry/&gt; (Updated July 2010). (Accessed August 2010).

9. American Cancer Society. Black Cohosh. American Cancer Society. &lt;http://www.cancer.org/Treatment/TreatmentsandSideEffects/Complementar yandAlternativeMedicine/HerbsVitaminsandMinerals/black-cohosh&gt; (Accessed August 2010).

10. Office of Dietary Supplements. National Institute of Health. &lt;http://ods.od.nih.gov/factsheets/BlackCohosh.asp&gt; (Updated November 21, 2008). (Accessed August 2010).

11. National Center for Complementary and Alternative Medicine. Black Cohosh. National Center for Complementary and Alternative Medicine. &lt;http://nccam.nih.gov/health/blackcohosh/&gt; (Updated August 28, 2009). (Accessed August 2010).

12. American Cancer Society. Bromelain. American Cancer Society. &lt;http://www.cancer.org/Treatment/TreatmentsandSideEffects/Complementar yandAlternativeMedicine/HerbsVitaminsandMinerals/bromelain&gt; (Accessed August 2010).

13. American Cancer Society. Calcium. American Cancer Society. &lt;http://www.cancer.org/Treatment/TreatmentsandSideEffects/Complementar yandAlternativeMedicine/HerbsVitaminsandMinerals/calcium&gt; (Accessed August 2010).

14. National Cancer Institute. National Cancer Institute FactSheet – Calcium and Cancer Prevention: Strengths and Limits of the Evidence. National Cancer Institute. &lt;http://www.cancer.gov/cancertopics/factsheet/prevention/calcium&gt; (Updated May 4, 2009). (Accessed August 2010).

15. American Cancer Society. Capsicum. American Cancer Society. &lt;http://www.cancer.org/Treatment/TreatmentsandSideEffects/Complementar yandAlternativeMedicine/HerbsVitaminsandMinerals/capsicum&gt; (Accessed August 2010).

16. Office of Dietary Supplements. Carnitine. National Institute of Health. &lt;http://ods.od.nih.gov/factsheets/carnitine.asp&gt; (Updated June 15, 2006). (Accessed August 2010).

17. American Cancer Society. Castor Oil. American Cancer Society. <http://www.cancer.org/Treatment/TreatmentsandSideEffects/Complementar yandAlternativeMedicine/ManualHealingandPhysicalTouch/castor-oil> (Accessed August 2010).

18. National Center for Complementary and Alternative Medicine. Chasteberry. National Center for Complementary and Alternative Medicine. <http://nccam.nih.gov/health/chasteberry/> (Updated July 2010). (Accessed August 2010).

19. National Center for Complementary and Alternative Medicine. Thunder God Vine. National Center for Complementary and Alternative Medicine. <http://nccam.nih.gov/health/tgvine/> (Updated July 2010). (Accessed August 2010).

20. American Cancer Society. Chlorella. American Cancer Society. <http://www.cancer.org/Treatment/TreatmentsandSideEffects/Complementar yandAlternativeMedicine/HerbsVitaminsandMinerals/chlorella> (Accessed August 2010).

21. American Cancer Society. Cloves. American Cancer Society. <http://www.cancer.org/Treatment/TreatmentsandSideEffects/Complementar yandAlternativeMedicine/HerbsVitaminsandMinerals/cloves> (Accessed August 2010).

22. American Cancer Society. Comfrey. American Cancer Society. <http://www.cancer.org/Treatment/TreatmentsandSideEffects/Complementar yandAlternativeMedicine/HerbsVitaminsandMinerals/comfrey> (Accessed August 2010).

23. American Cancer Society. Copper. American Cancer Society. <http://www.cancer.org/Treatment/TreatmentsandSideEffects/Complementar yandAlternativeMedicine/HerbsVitaminsandMinerals/comfrey> (Accessed August 2010).

24. American Cancer Society. Ellagic Acid. American Cancer Society. <http://www.cancer.org/Treatment/TreatmentsandSideEffects/Complementar yandAlternativeMedicine/DietandNutrition/ellagic-acid> (Accessed August 2010).

25. National Center for Complementary and Alternative Medicine. Feverfew. National Center for Complementary and Alternative Medicine. <http://nccam.nih.gov/health/feverfew/> (Updated July 2010). (Accessed August 2010).

26. American Cancer Society. Flaxseed. American Cancer Society. <http://www.cancer.org/Treatment/TreatmentsandSideEffects/Complementar yandAlternativeMedicine/HerbsVitaminsandMinerals/flaxseed> (Accessed August 2010).

27. National Center for Complementary and Alternative Medicine. Flaxseed and Flaxseed Oil. National Center for Complementary and Alternative Medicine. <http://nccam.nih.gov/health/flaxseed/> (Updated July 2010). (Accessed August 2010).

28. American Cancer Society. Folic Acid. American Cancer Society. <http://www.cancer.org/Treatment/TreatmentsandSideEffects/Complementar yandAlternativeMedicine/HerbsVitaminsandMinerals/folic-acid> (Accessed August 2010).

29. Office of Dietary Supplements. Dietary Supplement Fact Sheet: Folate. National Institute of Health. <http://ods.od.nih.gov/factsheets/folate.asp> (Updated April 15, 2009). (Accessed August 2010).

30. American Cancer Society. Garlic. American Cancer Society. <http://www.cancer.org/Treatment/TreatmentsandSideEffects/Complementar yandAlternativeMedicine/DietandNutrition/garlic> (Accessed August 2010).

31. National Cancer Institute. Garlic and Cancer: Questions and Answers. National Cancer Institute. <http://www.cancer.gov/cancertopics/factsheet/Prevention/garlic-and-cancer-prevention> (Updated January 22, 2008). (Accessed August 2010).

32. National Center for Complementary and Alternative Medicine. Garlic. National Center for Complementary and Alternative Medicine. <http://nccam.nih.gov/health/garlic/ataglance.htm> (Updated July 2010). (Accessed August 2010).

33. American Cancer Society. Ginger. American Cancer Society. <http://www.cancer.org/Treatment/TreatmentsandSideEffects/Complementar

yandAlternativeMedicine/HerbsVitaminsandMinerals/ginger> (Accessed August 2010).

34. National Center for Complementary and Alternative Medicine. Ginger. National Center for Complementary and Alternative Medicine. <http://nccam.nih.gov/health/ginger/> (Updated July 2010). (Accessed August 2010).

35. American Cancer Society. Ginkgo. American Cancer Society. <http://www.cancer.org/Treatment/TreatmentsandSideEffects/Complementar yandAlternativeMedicine/HerbsVitaminsandMinerals/ginkgo> (Accessed August 2010).

36. National Center for Complementary and Alternative Medicine. Ginkgo. National Center for Complementary and Alternative Medicine. <http://nccam.nih.gov/health/ginkgo/> (Updated July 2010). (Accessed August 2010).

37. American Cancer Society. Ginseng. American Cancer Society. <http://www.cancer.org/Treatment/TreatmentsandSideEffects/Complementar yandAlternativeMedicine/HerbsVitaminsandMinerals/ginseng> (Accessed August 2010).

38. National Center for Complementary and Alternative Medicine. Asian Ginseng. National Center for Complementary and Alternative Medicine. <http://nccam.nih.gov/health/asianginseng/> (Updated July 2010). (Accessed August 2010).

39. American Cancer Society. Gotu Kola. American Cancer Society. <http://www.cancer.org/Treatment/TreatmentsandSideEffects/Complementar yandAlternativeMedicine/HerbsVitaminsandMinerals/gotu-kola> (Accessed August 2010).

40. American Cancer Society. Grapes. American Cancer Society. <http://www.cancer.org/Treatment/TreatmentsandSideEffects/Complementar yandAlternativeMedicine/DietandNutrition/grapes> (Accessed August 2010).

41. National Center for Complementary and Alternative Medicine. Grape Seed Extract. National Center for Complementary and Alternative Medicine.

<http://nccam.nih.gov/health/grapeseed/> (Updated July 2010). (Accessed August 2010).

42.  American Cancer Society. Green Tea. American Cancer Society. <http://www.cancer.org/Treatment/TreatmentsandSideEffects/Complementar yandAlternativeMedicine/HerbsVitaminsandMinerals/green-tea> (Accessed August 2010).

43. National Cancer Institute. National Cancer Institute FactSheet –Tea and Cancer Prevention: Fact Sheet. National Cancer Institute. <http://www.cancer.gov/cancertopics/factsheet/prevention/tea> (Updated December 6. 2002). (Accessed August 2010).

44. National Center for Complementary and Alternative Medicine. Green Tea. National Center for Complementary and Alternative Medicine. <http://nccam.nih.gov/health/greentea/> (Updated July 2010). (Accessed August 2010).

45. National Center for Complementary and Alternative Medicine. Hawthorn. National Center for Complementary and Alternative Medicine. <http://nccam.nih.gov/health/hawthorn/> (Updated May 2008). (Accessed August 2010).

46. National Center for Complementary and Alternative Medicine. Horse Chestnut. National Center for Complementary and Alternative Medicine. <http://nccam.nih.gov/health/horsechestnut/> (Updated June 2008). (Accessed August 2010).

47. American Cancer Society. Indian Snakeroot. American Cancer Society. <http://www.cancer.org/Treatment/TreatmentsandSideEffects/Complementar yandAlternativeMedicine/HerbsVitaminsandMinerals/indian-snakeroot> (Accessed August 2010).

48.  American Cancer Society. Kava. American Cancer Society. <http://www.cancer.org/Treatment/TreatmentsandSideEffects/Complementar yandAlternativeMedicine/HerbsVitaminsandMinerals/kava> (Accessed August 2010).

49. National Center for Complementary and Alternative Medicine. Kava. National Center for Complementary and Alternative Medicine.

<http://nccam.nih.gov/health/kava/> (Updated July 2010). (Accessed August 2010).

50. National Center for Complementary and Alternative Medicine. Lavender. National Center for Complementary and Alternative Medicine. <http://nccam.nih.gov/health/lavender/> (Updated July 2010). (Accessed August 2010).

51. American Cancer Society. Licorice. American Cancer Society. <http://www.cancer.org/Treatment/TreatmentsandSideEffects/Complementar yandAlternativeMedicine/HerbsVitaminsandMinerals/licorice> (Accessed August 2010).

52. National Center for Complementary and Alternative Medicine. Licorice Root. National Center for Complementary and Alternative Medicine. <http://nccam.nih.gov/health/licoriceroot/> (Updated July 2010). (Accessed August 2010).

53. American Cancer Society. Lycopene. American Cancer Society. <http://www.cancer.org/Treatment/TreatmentsandSideEffects/Complementar yandAlternativeMedicine/DietandNutrition/lycopene> (Accessed August 2010).

54. American Cancer Society. Maitake Mushroom. American Cancer Society. <http://www.cancer.org/Treatment/TreatmentsandSideEffects/Complementar yandAlternativeMedicine/DietandNutrition/maitake-mushrooms> (Accessed August 2010).

55. American Cancer Society. Marijuana. American Cancer Society. <http://www.cancer.org/Treatment/TreatmentsandSideEffects/Complementar yandAlternativeMedicine/HerbsVitaminsandMinerals/marijuana> (Accessed August 2010).

56. National Cancer Institute. Marijuana Use in Supportive Care for Cancer Patients. National Cancer Institute. <http://www.cancer.gov/cancertopics/factsheet/Support/marijuana> (Updated December 12, 2000). (Accessed August 2010).

57. American Cancer Society. Melatonin. American Cancer Society. <http://www.cancer.org/Treatment/TreatmentsandSideEffects/Complementar

yandAlternativeMedicine/PharmacologicalandBiologicalTreatment/melatonin > (Accessed August 2010).

58. U.S. Department of Health and Human Services. Melatonin for Treatment of Sleep Disorders. Agency for Healthcare Research and Quality. <http://www.ahrq.gov/clinic/epcsums/melatsum.htm> (Updated November 2004). (Accessed August 2010).

59. American Cancer Society. Milk Thistle. American Cancer Society. <http://www.cancer.org/Treatment/TreatmentsandSideEffects/Complementar yandAlternativeMedicine/HerbsVitaminsandMinerals/milk-thistle> (Accessed August 2010).

60. National Cancer Institute. Questions and Answers About Milk Thistle. National Cancer Institute. <http://www.cancer.gov/cancertopics/pdq/cam/milkthistle/Patient/page2> (Updated June 18, 2010). (Accessed August 2010).

61. National Center for Complementary and Alternative Medicine. Milk Thistle. National Center for Complementary and Alternative Medicine. <http://nccam.nih.gov/health/milkthistle/ataglance.htm> (Updated July 2010). (Accessed August 2010).

62. American Cancer Society. Modified Citrus Pectin. American Cancer Society. <http://www.cancer.org/Treatment/TreatmentsandSideEffects/Complementar yandAlternativeMedicine/DietandNutrition/modified-citrus-pectin> (Accessed August 2010).

63. American Cancer Society. Noni Plant. American Cancer Society. <http://www.cancer.org/Treatment/TreatmentsandSideEffects/Complementar yandAlternativeMedicine/DietandNutrition/noni-plant> (Accessed August 2010).

64. National Center for Complementary and Alternative Medicine. Noni. National Center for Complementary and Alternative Medicine. <http://nccam.nih.gov/health/noni/> (Updated July 2010). (Accessed August 2010).

65. American Cancer Society. Omega-3 Fatty Acids. American Cancer Society. <http://www.cancer.org/Treatment/TreatmentsandSideEffects/Complementar

yandAlternativeMedicine/DietandNutrition/omega-3-fatty-acids> (Accessed August 2010).

66. U.S. National Library of Medicine and the National Institute of Health. Omega-3 Fatty Acids, Fish Oil, Alpha-Linolenic Acid. Medline Plus. <http://www.nlm.nih.gov/medlineplus/druginfo/natural/patient-fishoil.html> (Updated July 20, 2010). (Accessed August 2010).

67. U.S. Department of Health and Human Services. Omega-3 Fatty Acids, Effects of Cancer. Disorders. Agency for Healthcare Research and Quality. <http://www.ahrq.gov/clinic/tp/o3cantp.htm> (Updated January 2006). (Accessed August 2010).

68. National Center for Complementary and Alternative Medicine. Omega-3 Fatty Supplements: An Introduction. National Center for Complementary and Alternative Medicine. <http://nccam.nih.gov/health/omega3/introduction.htm> (Updated July 2009). (Accessed August 2010).

69. American Cancer Society. Peppermint. American Cancer Society. < http://www.cancer.org/Treatment/TreatmentsandSideEffects/Complementarya ndAlternativeMedicine/HerbsVitaminsandMinerals/peppermint > (Accessed August 2010).

70. National Center for Complementary and Alternative Medicine. Peppermint Oil. National Center for Complementary and Alternative Medicine. <http://nccam.nih.gov/health/peppermintoil/> (Updated July 2009). (Accessed August 2010).

71. American Cancer Society. Pokeweed. American Cancer Society. <http://www.cancer.org/Treatment/TreatmentsandSideEffects/Complementar yandAlternativeMedicine/HerbsVitaminsandMinerals/pokeweed> (Accessed August 2010).

72. American Cancer Society. Psyllium. American Cancer Society. <http://www.cancer.org/Treatment/TreatmentsandSideEffects/Complementar yandAlternativeMedicine/HerbsVitaminsandMinerals/psyllium> (Accessed August 2010).

73. American Cancer Society. Rabdosia Rubescens. American Cancer Society. <http://www.cancer.org/Treatment/TreatmentsandSideEffects/Complementar yandAlternativeMedicine/HerbsVitaminsandMinerals/rabdosia-rubescens> (Accessed August 2010).

74. American Cancer Society. Saw Palmetto. American Cancer Society. <http://www.cancer.org/Treatment/TreatmentsandSideEffects/Complementar yandAlternativeMedicine/HerbsVitaminsandMinerals/saw-palmetto> (Accessed August 2010).

75. National Center for Complementary and Alternative Medicine. Saw Palmetto. National Center for Complementary and Alternative Medicine. <http://nccam.nih.gov/health/palmetto/> (Updated July 2010). (Accessed August 2010).

76. American Cancer Society. Sea Vegetables. American Cancer Society. <http://www.cancer.org/Treatment/TreatmentsandSideEffects/Complementar yandAlternativeMedicine/DietandNutrition/sea-vegetables> (Accessed August 2010).

77. American Cancer Society. Shark Liver Oil. American Cancer Society. <http://www.cancer.org/Treatment/TreatmentsandSideEffects/Complementar yandAlternativeMedicine/PharmacologicalandBiologicalTreatment/shark-liver-oil> (Accessed August 2010).

78. American Cancer Society. Shiitake Mushroom. American Cancer Society. <http://www.cancer.org/Treatment/TreatmentsandSideEffects/Complementar yandAlternativeMedicine/DietandNutrition/shiitake-mushroom> (Accessed August 2010).

79. American Cancer Society. Soybean. American Cancer Society. <http://www.cancer.org/Treatment/TreatmentsandSideEffects/Complementar yandAlternativeMedicine/DietandNutrition/soybean> (Accessed August 2010).

80. U.S. Department of Health and Human Services. Effects of Soy on Health Outcome. Agency for Healthcare Research and Quality. <http://www.ahrq.gov/clinic/epcsums/soysum.htm> (Updated August 2005). (Accessed August 2010).

81. National Center for Complementary and Alternative Medicine. Soy. National Center for Complementary and Alternative Medicine. <http://nccam.nih.gov/health/soy/> (Updated July 2010). (Accessed August 2010).

82. American Cancer Society. St. John's Wort. American Cancer Society. <http://www.cancer.org/Treatment/TreatmentsandSideEffects/Complementar yandAlternativeMedicine/HerbsVitaminsandMinerals/st-johns-wort> (Accessed August 2010).

83. National Center for Complementary and Alternative Medicine. St. John's Wort. National Center for Complementary and Alternative Medicine. <http://nccam.nih.gov/health/stjohnswort/ataglance.htm> (Updated July 2010). (Accessed August 2010).

84. American Cancer Society. Tea Tree Oil. American Cancer Society. <http://www.cancer.org/Treatment/TreatmentsandSideEffects/Complementar yandAlternativeMedicine/HerbsVitaminsandMinerals/tea-tree-oil> (Accessed August 2010).

85. American Cancer Society. Turmeric. American Cancer Society. <http://www.cancer.org/Treatment/TreatmentsandSideEffects/Complementar yandAlternativeMedicine/HerbsVitaminsandMinerals/turmeric> (Accessed August 2010).

86. National Center for Complementary and Alternative Medicine. Tumeric. National Center for Complementary and Alternative Medicine. <http://nccam.nih.gov/health/turmeric/> (Updated July 2010). (Accessed August 2010).

87. American Cancer Society. Valerian. American Cancer Society. <http://www.cancer.org/Treatment/TreatmentsandSideEffects/Complementar yandAlternativeMedicine/HerbsVitaminsandMinerals/valerian> (Accessed August 2010).

88. National Center for Complementary and Alternative Medicine. Valerian. National Center for Complementary and Alternative Medicine. <http://nccam.nih.gov/health/valerian/> (Updated July 2010). (Accessed August 2010).

89. American Cancer Society. Vitamin C. American Cancer Society. <http://www.cancer.org/Treatment/TreatmentsandSideEffects/Complementar yandAlternativeMedicine/HerbsVitaminsandMinerals/vitamin-c> (Accessed August 2010).

90. American Cancer Society. Vitamin D. American Cancer Society. <http://www.cancer.org/Treatment/TreatmentsandSideEffects/Complementar yandAlternativeMedicine/HerbsVitaminsandMinerals/vitamin-d> (Accessed August 2010).

91. American Cancer Society. Apitherapy. American Cancer Society. <http://www.cancer.org/Treatment/TreatmentsandSideEffects/Complementar yandAlternativeMedicine/PharmacologicalandBiologicalTreatment/apitherapy > (Accessed August 2010).

92. National Cancer Institute. Questions and Answers About Aromatherapy. National Cancer Institute. <http://www.cancer.gov/cancertopics/pdq/cam/aromatherapy/Patient/page2> (Updated March 26, 2010). (Accessed August 2010).

93. American Cancer Society. Aromatherapy. American Cancer Society. <http://www.cancer.org/Treatment/TreatmentsandSideEffects/Complementar yandAlternativeMedicine/MindBodyandSpirit/aromatherapy> (Accessed August 2010).

94. American Cancer Society. Art Therapy. American Cancer Society. <http://www.cancer.org/Treatment/TreatmentsandSideEffects/Complementar yandAlternativeMedicine/MindBodyandSpirit/art-therapy> (Accessed August 2010).

95. American Cancer Society. Acupressure, Shiatsu, and Other Asian Bodywork. American Cancer Society. <http://www.cancer.org/Treatment/TreatmentsandSideEffects/Complementar yandAlternativeMedicine/ManualHealingandPhysicalTouch/acupressure-shiatsu-and-other-asian-bodywork> (Accessed August 2010).

96. American Cancer Society. Acupuncture. American Cancer Society. <http://www.cancer.org/Treatment/TreatmentsandSideEffects/Complementar

yandAlternativeMedicine/ManualHealingandPhysicalTouch/acupuncture> (Accessed August 2010).

97. National Cancer Institute. Questions and Answers About Acupuncture. National Cancer Institute. <http://www.cancer.gov/cancertopics/pdq/cam/acupuncture/Patient/page2> (Updated March 10, 2010) (Accessed August 2010).

98. National Center for Complementary and Alternative Medicine. Acupuncture: An Introduction. National Center for Complementary and Alternative Medicine. <http://nccam.nih.gov/health/acupuncture/introduction.htm> (Updated October 13, 2009). (Accessed August 2010).

99. National Center for Complementary and Alternative Medicine. Ayurvedic Medicine: An Introduction. National Center for Complementary and Alternative Medicine. <http://nccam.nih.gov/health/ayurveda/introduction.htm> (Updated July 2009). (Accessed August 2010).

100. American Cancer Society. Ayurveda. American Cancer Society. <http://www.cancer.org/Treatment/TreatmentsandSideEffects/Complementar yandAlternativeMedicine/MindBodyandSpirit/ayurveda> (Accessed August 2010).

101. American Cancer Society. Biofeedback. American Cancer Society. <http://www.cancer.org/Treatment/TreatmentsandSideEffects/Complementar yandAlternativeMedicine/MindBodyandSpirit/biofeedback> (Accessed August 2010).

102. American Cancer Society. Bodywork. American Cancer Society. <http://www.cancer.org/Treatment/TreatmentsandSideEffects/Complementar yandAlternativeMedicine/ManualHealingandPhysicalTouch/bodywork> (Accessed August 2010).

103. American Cancer Society. Breathwork. American Cancer Society. <http://www.cancer.org/Treatment/TreatmentsandSideEffects/Complementar yandAlternativeMedicine/MindBodyandSpirit/breathwork> (Accessed August 2010).

104. National Center for Complementary and Alternative Medicine. Complementary and Alternative Medicine Use and Children. National Center for Complementary and Alternative Medicine. <http://nccam.nih.gov/health/children/> (Updated April 2010). (Accessed August 2010).

105. American Cancer Society. Chinese Herbal Medicine. American Cancer Society. <http://www.cancer.org/Treatment/TreatmentsandSideEffects/Complementar yandAlternativeMedicine/HerbsVitaminsandMinerals/chinese-herbal-medicine> (Accessed August 2010).

106. American Cancer Society. Chiropractic. American Cancer Society. <http://www.cancer.org/Treatment/TreatmentsandSideEffects/Complementar yandAlternativeMedicine/ManualHealingandPhysicalTouch/chiropractic> (Accessed August 2010).

107. National Center for Complementary and Alternative Medicine. Chiropractic: An Introduction. National Center for Complementary and Alternative Medicine. <http://nccam.nih.gov/health/chiropractic/> (Updated November 11, 2009). (Accessed August 2010).

108. American Cancer Society. Dance Therapy. American Cancer Society. <http://www.cancer.org/Treatment/TreatmentsandSideEffects/Complementar yandAlternativeMedicine/MindBodyandSpirit/dance-therapy> (Accessed August 2010).

109. American Cancer Society. Electrodermal Screening. American Cancer Society. <http://www.cancer.org/Treatment/TreatmentsandSideEffects/Complementar yandAlternativeMedicine/ManualHealingandPhysicalTouch/electrodermal-screening> (Accessed August 2010).

110. American Cancer Society. Faith Healing. American Cancer Society. <http://www.cancer.org/Treatment/TreatmentsandSideEffects/Complementar yandAlternativeMedicine/MindBodyandSpirit/faith-healing> (Accessed August 2010).

111. American Cancer Society. Feng Shui. American Cancer Society. <http://www.cancer.org/Treatment/TreatmentsandSideEffects/Complementar

yandAlternativeMedicine/MindBodyandSpirit/feng-shui> (Accessed August 2010).

112. American Cancer Society. Humor Therapy. American Cancer Society. <http://www.cancer.org/Treatment/TreatmentsandSideEffects/Complementar yandAlternativeMedicine/MindBodyandSpirit/humor-therapy> (Accessed August 2010).

113. American Cancer Society. Hydrotherapy. American Cancer Society. <http://www.cancer.org/Treatment/TreatmentsandSideEffects/Complementar yandAlternativeMedicine/ManualHealingandPhysicalTouch/hydrotherapy> (Accessed August 2010).

114. American Cancer Society. Kampo. American Cancer Society. <http://www.cancer.org/Treatment/TreatmentsandSideEffects/Complementar yandAlternativeMedicine/HerbsVitaminsandMinerals/kampo> (Accessed August 2010).

115. American Cancer Society. Massage. American Cancer Society. <http://www.cancer.org/Treatment/TreatmentsandSideEffects/Complementar yandAlternativeMedicine/HerbsVitaminsandMinerals/massage?sitearea=ETO > (Accessed August 2010).

116. National Center for Complementary and Alternative Medicine. Massage Therapy: An Introduction. National Center for Complementary and Alternative Medicine. <http://nccam.nih.gov/health/massage/> (Updated June 2009). (Accessed August 2010).

117. American Cancer Society. Music Therapy. American Cancer Society. <http://www.cancer.org/Treatment/TreatmentsandSideEffects/Complementar yandAlternativeMedicine/MindBodyandSpirit/music-therapy> (Accessed August 2010).

118. American Cancer Society. Native American Healing. American Cancer Society. <http://www.cancer.org/Treatment/TreatmentsandSideEffects/Complementar yandAlternativeMedicine/MindBodyandSpirit/native-american-healing> (Accessed August 2010).

119. National Cancer Institute. National Cancer Institute FactSheet – Physical Activity and Cancer. National Cancer Institute. <http://www.cancer.gov/cancertopics/factsheet/prevention/physicalactivity> (Updated July 22, 2009). (Accessed August 2010).

120. American Cancer Society. Polarity Therapy. American Cancer Society. <http://www.cancer.org/Treatment/TreatmentsandSideEffects/Complementar yandAlternativeMedicine/ManualHealingandPhysicalTouch/polarity-therapy> (Accessed August 2010).

121. American Cancer Society. Qigong. American Cancer Society. <http://www.cancer.org/Treatment/TreatmentsandSideEffects/Complementar yandAlternativeMedicine/MindBodyandSpirit/qigong> (Accessed August 2010).

122. American Cancer Society. Reflexology. American Cancer Society. <http://www.cancer.org/Treatment/TreatmentsandSideEffects/Complementar yandAlternativeMedicine/ManualHealingandPhysicalTouch/reflexology> (Accessed August 2010).

123. American Cancer Society. Reiki. American Cancer Society. <http://www.cancer.org/Treatment/TreatmentsandSideEffects/Complementar yandAlternativeMedicine/ManualHealingandPhysicalTouch/reiki> (Accessed August 2010).

124. National Center for Complementary and Alternative Medicine. Reiki: An Introduction. National Center for Complementary and Alternative Medicine. <http://nccam.nih.gov/health/reiki/> (Updated July 2009). (Accessed August 2010).

125. American Cancer Society. Shamanism. American Cancer Society. <http://www.cancer.org/Treatment/TreatmentsandSideEffects/Complementar yandAlternativeMedicine/MindBodyandSpirit/shamanism> (Accessed August 2010).

126. American Cancer Society. Spirituality and Prayer. American Cancer Society. <http://www.cancer.org/Treatment/TreatmentsandSideEffects/Complementar yandAlternativeMedicine/MindBodyandSpirit/spirituality-and-prayer> (Accessed August 2010).

127. National Cancer Institute. Spirituality and Quality of Life. National Cancer Institute. <http://www.cancer.gov/cancertopics/pdq/supportivecare/spirituality/Patient/page2> (Updated March 6, 2009). (Accessed August 2010).

128. American Cancer Society. Tai Chi. American Cancer Society. <http://www.cancer.org/Treatment/TreatmentsandSideEffects/Complementar yandAlternativeMedicine/MindBodyandSpirit/tai-chi> (Accessed August 2010).

129. National Center for Complementary and Alternative Medicine. Tai Chi. National Center for Complementary and Alternative Medicine. <http://nccam.nih.gov/health/taichi/> (Updated November 1, 2008). (Accessed August 2010).

www.ingramcontent.com/pod-product-compliance
Lightning Source LLC
Chambersburg PA
CBHW021903170526
45157CB00005B/1939